Construction Worker Tools

By Laura Hamilton Waxman

BUMBA BOOKS™

LERNER PUBLICATIONS ◆ MINNEAPOLIS

Note to Educators

Throughout this book, you'll find critical-thinking questions. These can be used to engage young readers in thinking critically about the topic and in using the text and photos to do so.

Lerner Publications Company
A division of Lerner Publishing Group, Inc.
241 First Avenue North
Minneapolis, MN 55401 USA

For reading levels and more information, look up this title at www.lernerbooks.com.

Main body text set in Helvetica Textbook Com Roman 23/49.
Typeface provided by Linotype AG.

Library of Congress Cataloging-in-Publication Data

The Cataloging-in-Publication Data for *Construction Worker Tools* is on file at the Library of Congress.
ISBN 978-1-5415-5556-3 (lib. bdg.)
ISBN 978-1-5415-7348-2 (pbk.)
ISBN 978-1-5415-5647-8 (eb pdf)

Manufactured in the United States of America
1-46011-42928-10/30/2018

Table of Contents

Construction Workers Build

Construction workers use tools to build.

They build homes and buildings.

They build roads and bridges.

Workers wear special gear

to stay safe.

They wear hard hats.

They wear gloves and boots.

Many workers also

wear a tool belt.

It holds small tools.

A hammer fits in a tool belt.

It pounds nails.

A tape measure measures things.

What might a worker need to measure?

Workers use a saw to

cut wood.

Other saws cut stone

or metal.

Workers can make holes with a drill.

This tool can turn screws too.

Construction workers also

use big machines.

A bulldozer pushes dirt.

A digger scoops dirt from

the ground.

A dump truck carries things that workers need.

It holds sand, gravel, and dirt.

Why might workers need sand, gravel, or dirt?

Some tools break apart old roads or buildings. Then workers can build something new.

Construction Worker Tools

hard hat

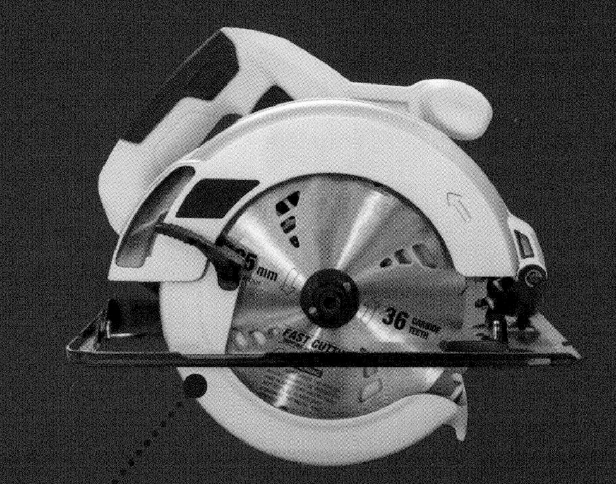

saw

hammer

digger

Picture Glossary

construction

the work of building something

machines

tools with moving parts

screws

small metal pieces that hold things together

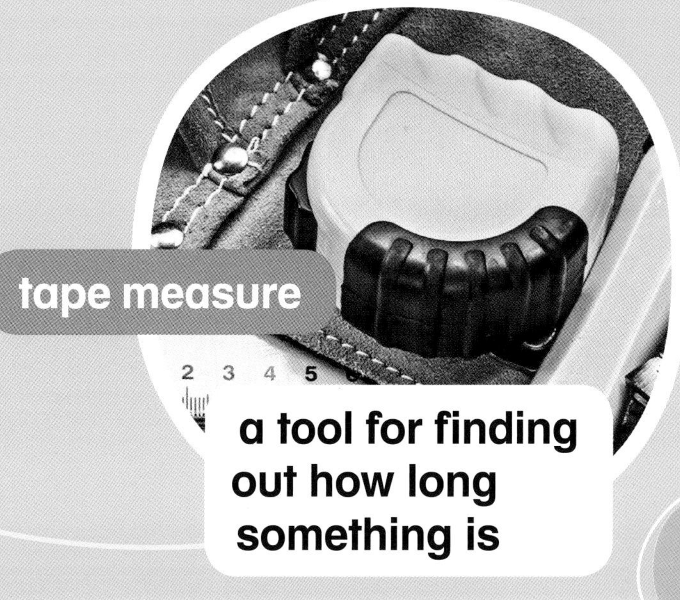

tape measure

a tool for finding out how long something is

Read More

Bowman, Chris. *Construction Workers*. Minneapolis: Bellwether Media, 2018.

Heos, Bridget. *Construction Workers in My Community*. Minneapolis: Lerner Publications, 2019.

Waldendorf, Kurt. *Hooray for Construction Workers!* Minneapolis: Lerner Publications, 2017.

Index

Photo Credits

Image credits: Hero Images/Getty Images, pp. 5, 23; Maskot/Getty Images, p. 6; Jiri Hera/Shutterstock.com, p. 9; mihalec/Shutterstock.com, pp. 11, 23; gpointstudio/Shutterstock.com, p. 13; XiXinXing/Shutterstock.com, p. 15; Dmitry Kalinovsky/Shutterstock.com, p. 16; TFoxFoto/Shutterstock.com, pp. 18, 23; thanaphoto/Shutterstock.com, p. 21; Alex Staroseltsev/Shutterstock.com, p. 22; MossStudio/Shutterstock.com, p. 22; Ton Bangkeaw/Shutterstock.com, p. 22; ESB Professional/Shutterstock.com, p. 23.

Cover Images: dslaven/Shutterstock.com; Dmitry Kalinovsky/Shutterstock.com.